A hay stacker in action at Wimborne Minster, Dorset, in 1951. Hay was swept direct from the field on to a pivoting platform that deposited its load on to the stack when hoisted by a tractor and cable.

HARVESTING MACHINERY

Roy Brigden

Shire Publications Ltd

CONTENTS

Printed in Great Britain by C. I. Thomas & Sons (Haverfordwest) Ltd, Press Buildings, Merlins Bridge, Haverfordwest, Dyfed SA61 1XF.

British Library Cataloguing in Publication Data available.

ACKNOWLEDGEMENTS
All photographs are the copyright of the Institute of Agricultural History and Museum of English Rural Life, Reading University. The drawing of the Roman *vallus* is by Godfrey Eke and the combine harvester on the front cover belongs to Mick Godfrey.

Cover: *A Massey Ferguson self-propelled combine harvester of 1956 at work in Oxfordshire, 1987.*

An artist's impression of the Roman vallus, based on the evidence of surviving stone reliefs in Europe.

Bell's reaper at work. (From the 'Encyclopaedia of Agriculture' by J. C. Loudon, second edition, 1831.)

HARVESTING CORN

Bringing in the harvest has always been the busiest time on the farm and attempts to speed up or mechanise the process are not new. The earliest identifiable harvesting device in Europe was the Roman *vallus* described by the writer Pliny in the first century AD as being in use on the larger estates of northern Gaul. Stone reliefs bearing depictions of the *vallus* have been found in the region, the best in what is now Luxembourg, and these confirm it to have been a wheeled frame with a row of long parallel fingers along its leading edge. It was pushed by a horse or mule into the standing corn, with the leading edge just below the height of the ears. The stalks were held in the narrow gap between the fingers as an operator walked alongside raking the ears off and back into a receptacle at the rear, from which they could be taken to the farmstead for threshing. This principle of removing only the ears or heads is not unlike that employed in the modern header type of combine harvester.

The *vallus* did not survive beyond the Roman Empire and no further progress was made on harvesting machines until the first quarter of the nineteenth century. A succession of inventions and patents, some more practical than others, appeared but none of them was built in great numbers or achieved lasting favour. In 1828, however, a reaping machine invented by the Reverend Patrick Bell of Scotland was tested by the Highland and Agricultural Society and was judged worthy of a £50 prize. It had a rotating reel at the front which drew the crop as it was cut on to a moving canvas, which in turn deposited it cleanly on the ground at the side, ready to be gathered by hand. The cutting was carried out by a double row of long pointed knives that worked in a scissor-like action just above ground level. The power for these operations came from the motion of the two large wheels of the machine as it was pushed by a pair of horses into the crop. It harvested at the rate of 1 acre (0.4 ha) an hour.

Three years of thought and experiment, from the days when he was still a student of divinity, had brought Bell's reaper to this level of public recognition. Although there were said to be ten of the machines in Scotland by 1832 and a few were sent to the USA, Australia and elsewhere, there was not yet production on any significant scale because there was little demand in Britain.

The Great Exhibition at the Crystal Palace in 1851 marked the beginning of a new phase. With more and more country people finding better paid employment in the towns and cities, many farmers were by now finding the cost of harvest labour expensive enough to warrant a closer look at the possibilities of mechanisation.

The McCormick reaper in the form in which it appeared at the Great Exhibition of 1851. Manual raking of the cut crop off the rear of the machine was still required at this stage.

Bell's reaper was not in the exhibition, but enormous interest was aroused by two machines developed in the United States, where severe labour shortfalls on vast new corn lands had spurred a more concerted attack on the harvesting problem.

Before bringing their machines to London, Cyrus Hall McCormick of Virginia and Obed Hussey, an ex-seaman based in Ohio, had vied with each other for a number of years over the primacy and relative merits of their machines. McCormick's reaper was first demonstrated publicly in 1831, although it did not receive a patent until the end of 1833, six months after Hussey's, and small-scale production began from a base on the family farm. Output rose rapidly during the 1840s as interest from the

Adapted or 'improved' versions of Hussey's reaper were produced by a number of English firms in the early 1850s. This one, dating from 1854, was by William Dray and Company of London.

The 'Omnium' self-raking reaper of 1879 by Samuelson and Company of Banbury. From his seated position the driver could cause the arms to pass over the platform without raking off the sheaf so that when cutting square corners the sheaf was deposited clear of the track of the horses on the next round. The price was £29.

prairie farmers of the Mid-western states accelerated and in 1847 a purpose-built McCormick works was established in Chicago, the new metropolis of the corn trade, capable of sending out five hundred reapers a year.

Hussey's machine could never quite match the McCormick in competition trials and did not reach the same level of volume production, even though in certain conditions it could do an excellent job. Both reapers were pulled rather than pushed by horses and required a second man in addition to the driver to rake the cut crop off the back and to the side of the platform. At the reaper trials held in conjunction with the Great Exhibition, the McCormick proved better equipped to tackle the damper crop conditions of England, but public interest was so great that trials were organised around the country and on occasion Hussey's machine was successful. Fired by the prospect of being able to cut up to 20 acres (8 ha) a day, demand for reapers from farmers was suddenly very heavy and a number of British firms immediately made arrangements to manufacture their own version of the McCormick or Hussey designs. Over 1500 machines are said to have been distributed by the end of 1853.

Such dramatic growth brought Bell's reaper out of obscurity and it was a Bell, in the form of an improved version being offered by Henry Crosskill of Beverley, Humberside, that triumphed over English versions of the American reapers in the trials organised by the Royal Agricultural Society in Gloucester in 1853. Patrick Bell's inventive foresight had finally been recognised, but even so take-up of the machine was comparatively slight, especially as it was being produced by only one of many firms now in the harvest machinery business.

The intense competition for supremacy between these firms ensured a steady flow of improvements in the efficiency and durability of reapers over the next twenty years. An early target with the American-style machines was to eliminate the human labour required to rake the crop off the platform. An early solution, introduced by Burgess and Key of Brentwood, Essex, in 1854, used an Archimedean screw for the purpose, but most attention focused on the incorporation of a mechanical raking arm. At first this was a separate unit which was operated through gearing from the main drive and mounted behind the reel to the rear of the machine. The crucial

A self-raking or sail reaper at work in about 1900.

development, however, patented by another American, Owen Dorsey, in 1856, came with the replacement of the reel by a cluster of rake arms, rotating on a vertical axis, both to draw the crop on to the cutter bar and then to sweep it back off the table in a continuous sequence. This quickly established itself as the standard arrangement for the self-raking reaper and the first British version was built by Samuelson and Company of Banbury, Oxfordshire, in the early 1860s. Many thousands were in use by the end of the century.

The distinctive 'Albion' binder by Harrison, McGregor and Company of Leigh in Lancashire first went into production in 1894 and many have been preserved. This is the Number 2 version of 1908. Each sheaf was tied in the open knotter mechanism on the right before being released to the ground.

A binder drawn by three horses at work in Berkshire in the 1940s. The sheaves were gathered together into stooks for drying before subsequent collection and stacking.

To keep pace with one of these reapers, four or five people were still needed to follow behind, gather the corn into sheaves and arrange them into stooks. A considerable reduction in manpower could be achieved by having two people riding on the reaper itself to tie into sheaves the corn that was delivered to them by a moving canvas table and then to drop the sheaves off the back. Charles Marsh of Illinois marketed the idea and, from the appearance of his first 'harvester' in 1858, demand in the USA rapidly grew, particularly when labour shortages induced by the Civil War (1861-5) became acute.

Such was the speed of American advance that in Britain there was almost no time to scrutinise the harvester before it was overtaken by a new generation of machines with automatic sheaf-binding systems. The first series in the early 1870s relied on wire-tying mechanisms, but difficulties associated with small pieces of wire passing through to cattle feed and processing machinery made it important to find a safer substitute. By 1880 the problem had been solved by switching to twine made from a mixture of sisal and Manila hemp. Within a few years twine binders, whether McCormick imports or local copies by firms such as Bamlett, Samuelson, Howard and Hornsby, were being mass-produced to supply a huge British market.

At the close of the nineteenth century more than three-quarters of the corn crop was being harvested by machine. Improvements to binders continued to appear but the essential features of those in use after the Second World War, even though pulled by tractors rather than horses, were little altered from their late nineteenth-century predecessors. The number of binders in Great Britain peaked in 1950 at 150,000, but thereafter began a steady decline as the combine harvester began to be adopted more widely.

A small threshing machine of 1847 housed in a barn and powered by a two-horse gear to which it is connected through shafting. Manufactured by Barrett, Exall and Andrewes of Reading, it had an advertised capacity of 15 quarters (190 kg) of wheat per day and cost £32.55.

A section through a portable threshing machine of the 1840s built by Richard Garrett of Leiston, Suffolk. The drum and concave, where threshing takes place, are at the top; the straw passes out at the rear; in the lower section, grain is separated from chaff using the fan positioned between the rear wheels.

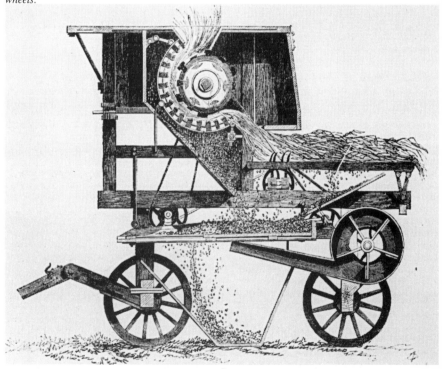

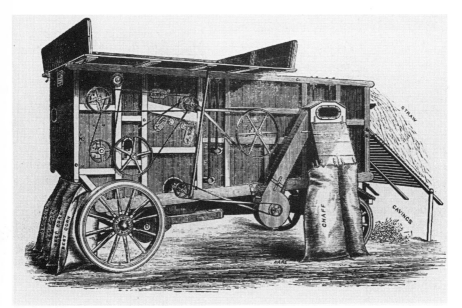

A portable finishing threshing machine of 1876 by Charles Burrell and Sons of Thetford, Norfolk, which delivered graded and cleaned corn direct into sacks ready for market. It is a double-blast machine, which means that the grain passes through two separate screening or dressing machines, each of which is assisted by the draught from a rotary fan.

THRESHING MACHINES

The mechanised processing of corn to separate grain from straw pre-dated the general use of harvesting equipment and relied far more heavily on British development work. The first effective threshing machine was built in 1786 by Andrew Meikle, a Scottish millwright, and worked on the principle of feeding the crop between two grooved rollers into the path of a revolving drum that used beater bars attached to its rim to knock out the grain. An outer casing or concave delivered both grain and straw on to a screen below, where the two were separated. This was the prelude to a quite rapid spread of similar machines over the next thirty years, mostly on larger farms in southern Scotland and the far north-eastern counties of England, where labour for hand threshing was at the time comparatively scarce and therefore expensive.

The earliest threshing machines were cumbersome pieces of equipment installed as fixtures within barns and usually driven by horse or water power, although stationary steam engines made their first appearance around 1800. In the more southern English counties the development of a simpler version, capable of being transported for hire from farm to farm, brought increased interest because it enabled farmers to take better advantage of dearer early-season grain prices at little cost to themselves. Here also the mechanism followed a slightly different evolutionary path, relying rather more on a rubbing than on a beating action and thus doing less damage to the straw, with the grain falling directly through the casing mesh in a preliminary separation. Further progress was temporarily halted when years of agricultural and rural depression culminated in 1830 in sporadic outbreaks of unrest in southern England. The threshing machine, popularly regarded as a destroyer of labour, took the brunt of the frustration and several hundred were destroyed.

As stability returned, so did demand,

9

A threshing scene in the early 1900s. A portable steam engine by Davey, Paxman and Company of Colchester, Essex, is driving a threshing machine by Nalder and Nalder of Challow, Oxfordshire with a straw elevator attached at the rear.

Clayton and Shuttleworth of Lincoln produced over twenty thousand threshing machines between 1842 and 1885 alone. They also sold large numbers of elevators, for use on both straw and hay, which were driven by a single-horse gear. As the stack grew in height, so the elevator was raised by means of a hand crank at the front.

A straw-trussing machine from the 1885 catalogue of Marshall, Sons and Company Limited of Gainsborough, Lincolnshire. It was positioned behind the threshing machine and linked to it by a drive chain. Threshed straw passed directly into the trusser hopper to be compressed into a sheaf and bound with twine bands.

and many improvements in the design and efficiency of threshing machines were made by the growing number of makers. From the 1850s most attention was directed at the perfection of mobile machines powered by portable steam engines, and then increasingly by self-propelled ones. These threshing sets, operated in large numbers by firms of travelling contractors, together with the smaller horse-powered and even hand-powered machines, saw the virtual elimination of the flail from farms in the last quarter of the century.

The threshed straw was mechanically handled into the stack by positioning an elevator at the end of the machine. After 1914 the elevator was replaced by a trusser if the straw was to be taken elsewhere. This used a double knotting mechanism, similar to that employed on binders, to deliver bound trusses ready for transport.

WINNOWING MACHINES

The more substantial of the early threshing machines were equipped with winnowing fans to separate the grain from the chaff. This principle of using the draught from a rotating fan to blow out chaff originated in China at least two thousand years ago and was brought to mainland Europe in the seventeenth century through improved trading links. The idea was not adopted immediately in Britain but early in the second half of the eighteenth century the basic format of the hand-cranked winnower was well established. With the addition of shaking sieves operated through cranks from the main drive, the machine was also able to dress the grain by grading it and removing dirt, small stones and seeds. In 1848 the first combined threshing and dressing machine was introduced, followed in the next decade by finishing machines which incorporated up to three separate fanning and sifting processes to clean and grade the grain ready for market. Nevertheless, there was still a place for the hand-powered dresser and it remained common on farms into the twentieth century because it produced a better marketable sample than the output of less efficient threshing machines and provided an effective means of checking that seed was pure and free of weeds before sowing.

11

A hand-powered dressing machine by Robert Boby of Bury St Edmunds, Suffolk, 1875. The fan blew off the chaff as with an ordinary winnowing machine but in addition the grain passed through riddles to divide off weed seeds and then along a self-cleaning screen to separate the best corn from the 'tail' or 'seconds'.

Tractors were commonly used to drive threshing machines in the 1940s. Here is a Fordson operating a Marshall steel-framed thresher to which is attached a stationary baler; Nottinghamshire, 1945.

Drawn by an army of horses, this combine harvester is of the type that began to appear on the huge corn farms of California in the early 1880s.

COMBINE HARVESTERS

The combine harvester brings harvesting and threshing together into a single field operation with consequent savings in time and labour. It first appeared in North America as early as the mid 1830s when Hiram Moore experimented in Michigan with a number of giant machines. With 15 foot (4.5 metre) cutter bars and pulled by twenty or more horses, these were capable of harvesting 20 acres (8 ha) a day. In Australia an extreme scarcity of harvest labour led to the emergence in 1843 of a new type of device known as the stripper or locomotive thresher. As with the Roman *vallus*, a long comb entered the crop just below head height but, in addition, a rotating beater bar removed the heads, carried them over a threshing concave and deposited the grain in a box at the rear. Winnowing remained a separate operation until combined into the stripper-harvester by Hugh McKay of Victoria in 1885. Marketed under the Sunshine Harvester name, these machines had become widely used in Australia by the early twentieth century and many were exported.

In the last quarter of the nineteenth century, California was the principal home of the combine, for here the warm dry climate provided ideal operating conditions and many of the new cereal farms were so vast that they could be worked only if the most was made of mechanisation. Amongst many developments was the first self-propelled combine in 1886, which, with its straw-fired steam engine, had a maximum output of 100 acres (40 ha) a day. In 1890 combines accounted for over 1.5 million acres (600,000 ha) of corn in the state and most of them were manufactured by the Holt Brothers or Daniel Best, the two leading companies.

Physical geography, economic factors and the need to conserve straw for subsequent livestock use made such machines impractical for use in Britain. Moreover, in a cooler and less settled climate combined grain would normally be too damp to store safely without first being artificially dried, thus bringing in an additional cost element. For these reasons standard procedure continued to depend on binding sheaves, stooking them in the field to assist the natural drying process, building stacks and then later threshing out the grain at a convenient time in the autumn or winter.

During the 1920s and 1930s lighter,

13

Because careless work in feeding the threshing machine could and did lead to gruesome accidents, optional self-feeding devices began to appear from the 1870s. In this Ransomes example of 1886, the cut sheaves were loaded into a hopper and then a revolving barrel ensured an even flow into the machine.

cheaper combines were developed for smaller farms of the British type. This was helped by improvements that had been made to the internal combustion engine, together with the introduction of power take-off (PTO) to tractors in 1924 and of economical construction techniques using angle-iron and sheet metal. The early breakthroughs again took place in North America, where the Canadian Massey Harris company, for example, in 1937 followed up a series of successful models with the Number 15, which had an 8 foot (2.4 metre) cut and was mounted on rubber tyres and PTO-driven. In 1938 they produced the very popular small-farm Clipper combine, which had its threshing section positioned directly behind the reel and cutter bar so that the crop passed straight through the

A British-built Clayton and Shuttleworth combine harvester at work on an Oxfordshire farm in 1932. It was powered by an auxiliary engine, had a 16 foot (488 cm) cut and required three men to operate. Few of these machines were built and none has survived.

The International Harvester Company was formed in 1902 through the merger of five leading American firms, including McCormick. Its first generation of combine harvesters, produced from 1914, went through successive improvements and here is one on a Suffolk farm in 1940.

machine. In the United States the biggest-selling small combine of this period was the Allis Chalmers Model 60 All Crop, a machine with a 5 foot (1.5 metre) cut and capable of being used with an ordinary tractor on a wide variety of crops. Three years after its introduction in 1934 annual production was running at over ten thousand units.

Although the first British-built combine was by the Lincoln firm of Clayton and Shuttleworth and appeared in 1928, it was this new generation of American and Canadian machines that began to attract small numbers of innovative farmers during the 1930s. About a hundred had been imported by 1939 and had demonstrated their ability to deal satisfactorily with European crop conditions and climate. The onset of the Second World War brought labour shortages, an increasing corn acreage and higher grain prices, all of which helped to stimulate a dramatic rise in the number of combines to an estimated 950 in 1942 and 2500 in 1944. After the war the trend was similar. Massey-Harris opened its factory at Kilmarnock, Strathclyde, in 1949, the output

of which contributed to doubling the number of combines in use on British farms between 1948 and 1950 to more than ten thousand; by 1960 the total had exceeded fifty thousand.

The cheapest combines of the war years and immediately after were those with power take-off drive from the pulling tractor. A more constant running of the threshing drum, which was vital for best results, was obtainable if the combine mechanism was driven by its own auxiliary engine while still relying on a tractor for haulage. Self-propelled machines became available in the early 1940s and, in spite of the higher initial cost, became steadily more popular because they were easier and more convenient to operate and more manoeuvrable in small fields. The work rate of many of these models depended primarily on the width of cutter bar, so the wartime figures ranged from 1 acre (0.4 ha) per hour for a machine with a 5 foot (1.5 metre) cutter to up to 2 acres (0.8 ha) per hour for one with a 12 foot (3.6 metre) cutter.

Initial fears that combines would oblige

15

The International Harvester Company's small tractor-drawn combines, like this Number 42 with a 4 foot (122 cm) cut photographed in 1941, proved popular in the United States and were suitable for British conditions.

An Allis-Chalmers Model 60 All Crop Harvester near Daventry, Northamptonshire, in 1947. These American machines brought combining to many British farms in the 1940s and 1950s at moderate cost. They were available either with a bagging platform or with a grain bin for bulk unloading.

A Sunshine Auto Header under trial in England in 1941. This was an Australian machine developed by the H. V. McKay Company from 1923. It had a 12 foot (366 cm) cut, taking the corn just below the ear, and was the first commercial combine to incorporate the threshing apparatus directly behind the cutter.

farmers to dispose of all their grain on a glutted market immediately after the harvest, when prices were at their lowest, led to the development of on-farm drying and storage systems that enabled the crop to be held for release at a more appropriate and remunerative time. For storage to be viable, the corn had to be harvested at the stage of dead ripeness, rather later than normal with a binder, and then its moisture content was reduced from around 20 per cent down to 14 per cent. A common type of dryer in the 1940s was a vertical structure in which a descending flow of wet grain had warm air and gases drawn through it from a furnace. It was then removed by conveyor to stores, which at this time might have taken the form of pre-cast concrete silos or of wooden bins constructed within a barn.

Massey-Harris Number 21 self-propelled combines at South Creake, Norfolk, in 1943. This model first appeared in 1940 and with its 12 foot (366 cm) cut was ideal for the larger British farms.

A Massey-Harris 21 at Weasenham, Norfolk, in 1954 unloading its grain tank.

Massey-Harris-Ferguson Limited introduced its version of the mounted combine in 1956. It could be fitted around a Ferguson tractor in about fifteen minutes to provide a highly manoeuvrable self-propelled combine with 7 foot (229 cm) cut at reduced cost.

A 'balance-draught' grass mower by Samuelson and Company, 1879. The drive for the reciprocating cutter blade was taken from the rim gear of the land wheels. When travelling, the whole cutter bar was folded upright to reduce the machine's width.

HAYMAKING

Hay rakes and some other horse-drawn implements for the improvement of the otherwise immensely labour-intensive process of haymaking were introduced in a small way from the beginning of the nineteenth century. The pace of change quickened very noticeably during the 1850s, when a practical and efficient grass-mowing machine was developed, using many of the same basic cutting principles that had first been tried on the reaper. Indeed, such were the similarities that some dual-purpose machines were offered which were capable, after only minor adjustments, of cutting both corn and grass. During the last quarter of the century mowing machines became widespread on British farms once design improvements had done much to make them lighter of draught, harder-wearing and effective under more uneven field conditions.

Following the mower's success, interest turned again to the subsequent stages of haymaking. A simple form of tedding machine, for tossing the cut grass to assist the drying process, had been in existence since at least 1815 and was now, in the second half of the century, improved and taken up in large numbers. In 1875 curved hoods were added in front of the revolving rakes and put a stop to the constant showering of the driver with grass. These machines remained popular into the twentieth century, especially where growth was exceptionally thick and heavy swaths of grass were required to be broken and scattered.

For ordinary work the tedder lost ground to a new device, the swath turner, once it reached a reliable form in 1896. This was not only much gentler in its treatment of the crop but was also able to invert the swath completely, usually on the second day after mowing, and deposit it on the now dry strip adjacent. A further innovation in the early twentieth century was the side-delivery rake, which collected two swaths together into a single windrow in order to complete the

A Paragon Number 2 combined mower and reaper by R. Hornsby of Grantham, Lincolnshire, 1902. It required two operators, one to drive and the other to rake the cut crop off the back of the machine.

Blackstone and Company of Stamford, Lincolnshire, claimed to have made over 23,000 haymakers by 1892. This is the prize-winning Taunton hooded haymaker with four sets of flyers on a steel axle. A ratchet escapement allowed the flyers to continue spinning when the wheels slowed or stopped in order to prevent clogging with hay.

A swath turner by Ransomes, Sims and Jefferies of Ipswich, introduced in 1896. Two sets of rotating flyers operated on a pair of swaths at the same time to turn the grass over gently on to the dry ground between the rows. With one horse, 30 acres (12 ha) a day could be worked.

drying before collection began. The process was completed in 1913 with the appearance of a combined machine for the three operations of tedding, swath turning and side raking.

The hay loader was a labour-saving device that found some favour on larger farms from the 1920s onwards. It worked by picking up hay direct from the windrow and hoisting it up a ladder of reciprocating rake bars into a cart or wagon in front. Once at the stack, the wagon could be unloaded with mechanic-al assistance either from an elevator, which had been available in both steam-powered and horse-driven form since the early 1870s, or from a grab and portable gantry that used ropes and a horse to provide lift.

Carting could be eliminated if stacks were built at the edge of the field rather than some distance away at the farmstead. Hay sweeps drawn by one or two horses were then all that was required for collecting. In the 1930s it was not uncommon for wooden sweeps to be fitted to

Bamfords of Uttoxeter, Staffordshire, introduced their first side-delivery rake for putting up hay into windows in 1908. This is the version of 1914.

Bamfords were pioneers of the combined side rake, swath turner and tedder, with the first model appearing in 1913. Here is a 1920s version in use in 1940.

A Monarch hay loader of the 1930s by L. R. Knapp and Company of Clanfield, Oxfordshire. Chain drive from both wheels provided the power for the cranks that operated the rake arms.

A Paragon horse hay fork from the 1927 catalogue of G. C. Ogle and Sons of Ripley, Derbyshire. By means of a pulley system, a horse drawing a rope away from the pole raised each load of hay to the top of the stack.

A typical steel-toothed horse hay rake of the late nineteenth century. This manual example by Blackstone and Company, 1892, required the lever to be pushed to raise the teeth and release the load but automatic versions were also available.

the front end of powerful old, and otherwise scrappable, motor cars to provide a quick and easy means of handling hay, which was so often vital in unsettled weather. As tractors became more popular, similar methods were tried on them, with the later refinement of the use of hydraulics to lift the sweep clear of the ground.

In the years immediately following the Second World War the automatic pick-up baler offered even greater labour savings at harvest time, as well as subsequently, for bales were easier to handle than loose hay when required for feeding stock. The commonest types in the 1950s were driven either by an auxiliary engine or from the tractor's power take-off. They picked up the hay directly from the windrow and rammed it into the size of bale required before tying it lengthwise with twine. Output could be as much

A wooden hay sweep attached to the front of a modified old motor car in the 1930s.

as 6 tons per hour on bales weighing around 50 pounds (22.5 kg) each, but care was always needed to ensure that they were not too tightly compressed, especially when affected by damp, for then the hay was likely to overheat and go mouldy.

A stationary baler by Powell and Company of St Helens, Merseyside. These were popular in the 1930s and 1940s for baling hay supplied to them direct by tractor sweeps in the field.

Clyde Higgs, a Warwickshire farmer, was a leading pioneer of the grass-drying process in the 1930s. Here is a batch dryer with its two drying trays being demonstrated on his farm in 1936.

GRASS DRYING

Other ways of conserving grass for winter livestock feed have been devised in the twentieth century and require additional types of specialised equipment. In the mid 1920s research demonstrated that when young short grass was cut and then artificially dried it was much richer in nutrients than hay made by conventional means. This led to a moderate, though by no means widespread, growth of on-farm drying systems that continued into the 1950s. The simplest form of dryer drew hot gases from a furnace successively across two large trays filled with grass to a depth of 18 inches (457mm), with each batch taking about 25 minutes to dry.

A Wilder Cutlift close-cropping grass and raising it into the specially designed trailer for transfer to a dryer, in the late 1930s.

A Bentall Cutter Blower chopping up sugar-beet tops and wheat straw before loading them into a concrete sectional silo, 1941.

More sophisticated versions produced a continuous stream of dried material by blowing hot air through the grass as it progressed along the machine on a conveyor. The whole process, from cutting a batch of grass to baling it from the dryer, could be over in an hour, so that inclement weather was less of a threat, and lower labour charges, together with the improved quality of the feed, helped offset the extra plant and fuel costs.

New equipment was also needed out in the field to cut the grass when it was only 4 or 5 inches (100-130 mm) high and transport it back to the dryer at the farmstead. A pioneer British example was the Cutlift, introduced by John Wilder Limited of Reading in 1935, which shaved the grass close to the ground and then, in one operation, elevated it into a trailer towed at the rear. The same firm's chopper-loader of 1953 used an auxiliary engine to chop up longer grass to the requisite size before lifting it between rubber-covered conveyors.

SILAGE MAKING

After something of a false start in the 1880s, silage making in Britain slowly re-established itself from the 1920s and has grown enormously in popularity since 1945. The key element is a carefully controlled fermentation process, from which air is excluded and which is applicable to a wide range of green crops in addition to grass and results in a highly nutritious livestock feed. Following American experience, tower silos were often preferred before the Second World War, but thereafter the less complicated walled clamp with an airtight seal provided by plastic sheeting weighed down by numerous old rubber tyres progressively became the commonest form.

Where the crop had to be conveyed only a short distance to the clamp, a buckrake mounted on the three-point linkage at the back of a tractor proved to be a cheap and easy method. Otherwise more substantial field machinery was needed, like the green-crop loader which used a pick-up cylinder and elevating device to collect the previously mown crop out of the swath or windrow. Combined equipment of the Wilder Cutlift type, which could cut and perhaps chop the crop as well, proved its worth on silage during the 1950s. Subsequent developments followed the American form of forage harvester which used a very fast rotary cutting action linked to an impelling device to shoot the crop through ducting into a high-sided and partly roofed trailer travelling behind or alongside.

26

A surface silage clamp with a top seal of thatch at Dunsfold, Surrey, in 1951.

An Allis-Chalmers forage harvester that was introduced in 1950. The cutting height for silage could be varied from 1 to 14 inches (25 to 355 mm) and the maximum capacity was 10 tons of grass per hour delivered via the chute into an accompanying trailer.

The Ransomes new potato digger of 1910. Its hanging forks worked in parallel with one another and brushed the potatoes to one side with the minimum of scattering or bruising. Each row then had to be picked up by hand and at this rate about 4 acres (1.6 ha) a day could be completed.

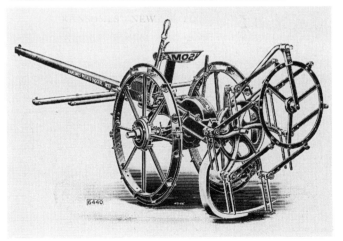

A potato harvester by Victor Nilssom of Sweden under trial in England, 1952. The crop was lifted by a four-bladed share and deposited into the sloping drum. The potatoes dropped on to a side-delivery elevator while the haulm carried on to exit at the back.

HARVESTING ROOT CROPS

The mechanisation of harvesting potato and sugar-beet crops was greatly stimulated from the 1930s when increases in both acreage and yield stretched more laborious methods to the limit. The area under potatoes in Britain, for example, rose from 400,000 acres (160,000 ha) in 1885 to over 1 million acres (400,000 ha) immediately following the Second World War and total production rose from under 2 million to over 8 million tons. Little sugar beet was grown until the building of the first British processing factory in Norfolk in 1912 and the deliberate encouragement of home-produced sugar. This continued during the 1920s, particularly in East Anglia, as a means of offsetting the depression amongst cereal growers. Consequently the total acreage rose from 8000 acres (3200 ha) in 1921 to a little over 400,000 acres (160,000 ha) during the Second World War, at which point it stabilised, and yields over the same period increased from 65,000 to almost 4 million tons. For both potatoes and sugar beet complete harvesters already existed before the outbreak of war, often with a continental origin and often in not much more than experimental form, but it was not until wartime restrictions on allocation of materials and production were relaxed in the late 1940s that their numbers began to increase significantly.

The simplest device for lifting potatoes is a plough-like implement with a flat share and, instead of a mouldboard, a row of angled prongs which bring the tubers to the surface and separate them from the soil. These first appeared in the middle of the nineteenth century but two-row tractor versions were still available over a century later partly because they always worked to some extent, even when adverse soil and weather conditions might defeat a more sophisticated machine. The more usual spinner principle was introduced in the 1850s, although it was another thirty years before it had progressed sufficiently to work well, without bruising or damaging the crop, in a variety of soils. The principal feature was a wide share that passed beneath the crop to loosen the soil, followed by forks rotating in either a vertical or an inclined plane which left the potatoes on the surface and a little to one side. Tractor-drawn versions of the 1950s were driven either from the power take-off or from the machine's own land wheels and could incorporate a second auxiliary spinner to make collection of the crop easier.

The next stage in mechanisation was the elevator digger, which was taken up in growing numbers from the Second World War. A broad inclined share raised the crop on to an endless chain web through which soil and small stones dropped as the potatoes were carried backwards and off the rear of the machine for manual picking. From this it was only a small progression to the complete harvester with its additional screening capacity for further removal of extraneous material before delivering the crop by conveyor direct into a trailer or on to an integrated bagging platform. From just a handful at the end of the war,

five hundred potato harvesters were in use by 1950 and over 2500 a decade later.

For sugar-beet harvesters the story is similar, with fewer than one thousand on British farms in 1950, but almost ten thousand in 1960. The build-up, through progressive mechanisation, of the component operations also followed a similar pattern. It began with horse-drawn, and subsequently tractor-drawn, beet lifters which left the crop on the ground ready for topping and collection by hand. Then came the automatic topping machine relying on a feeler device running over the beet in advance of the knife to control the depth of cut for beets of different size. First-generation complete harvesters were driven from the tractor power take-off and were arranged with the topping mechanism at the front, followed by lifting shares which raised the beet on to endless web conveyors to clean them of clods and load them on to a trailer.

A one-row potato digger by the John Deere Company of Moline, Illinois, working at Shillingford, Oxfordshire, in 1941. Driven by the tractor's power take-off, the machine raised the potatoes on to an agitating elevator to loosen the dirt before delivering them off the back for manual picking.

FURTHER READING

Bond, J. R. *Farm Implements and Machinery*. Benn Brothers, 1923.
Culpin, C. *Farm Machinery*. Granada Publishing, St Albans, Hertfordshire, 1938, and subsequent editions.
Fussell, G. E. *The Farmer's Tools 1500-1900*. Bloomsbury Books, 1985. First published by Andrew Melrose Limited, 1952.
Orr, J. *Machinery on the Farm*. Blackie, 1949.
Quick and Buchele. *The Grain Harvesters*. American Society of Agricultural Engineers, Michigan, 1978.

PLACES TO VISIT

Intending visitors are advised to establish dates and hours of opening before making a special journey.

Acton Scott Working Farm Museum, Wenlock Lodge, Acton Scott, Church Stretton, Shropshire SY6 6QN. Telephone: 06946 306 or 307.

Ashwell Village Museum, Swan Street, Ashwell, Baldock, Hertfordshire SG7 5NY. Telephone: 046274 2155.

Atholl Country Collection, The Old School, Blair Atholl, Perthshire PH18 5TT. Telephone: 079681 232.

Barleylands Farm Museum, Barleylands Road, Billericay, Essex CM11 2UD. Telephone: 0268 282090.

Beamish: The North of England Open Air Museum, Beamish Hall, Beamish, Stanley, County Durham DH9 0RG. Telephone: 0207 231811.

Beck Isle Museum of Rural Life, Beck Isle, Pickering, North Yorkshire YO18 8DU. Telephone: 0751 73653.

Bewdley Museum, The Shambles, Load Street, Bewdley, Worcestershire DY12 2AE. Telephone: 0299 403573.

Bickleigh Mill Farm, Bickleigh, Tiverton, Devon. Telephone: 08845 419.

Bicton Park Countryside Collection, East Budleigh, Budleigh Salterton, Devon EX9 7DP. Telephone: 0395 68465.

Bodmin Farm Park, Fletchers Bridge, Bodmin, Cornwall. Telephone: 0208 2074.

Brattle Farm Museum, Brattle Farm, Staplehurst, Tonbridge, Kent TN12 0HE. Telephone: 0580 89122.

Breamore Countryside Museum, Breamore House, Breamore, Fordingbridge, Hampshire SP6 2DF. Telephone: 0725 22270.

Bygones at Holkham, Holkham Park, Wells-next-the-Sea, Norfolk. Telephone: 0328 710806 or 710277.

Cambridge and County Folk Museum, 2/3 Castle Street, Cambridge CB3 0AQ. Telephone: 0223 355159.

Castle Cary Museum, The Market House, Castle Cary, Somerset BA7 7AA. Telephone: 0963 50277.

Church Farm Museum, Church Road South, Skegness, Lincolnshire. Telephone: 0754 66658.

Clitheroe Castle Museum, Castle Hill, Clitheroe, Lancashire BB7 1BA. Telephone: 0200 24635.

Cogges Farm Museum, Church Lane, Cogges, Witney, Oxfordshire OX8 6LA. Telephone: 0993 72602.

Cornish Museum, Lelant Model Park, Lelant, St Ives, Cornwall. Telephone: 0736 752676.

Cotswold Countryside Collection, Northleach, Cheltenham, Gloucestershire GL54 3JH. Telephone: (summer) 0451 60715, (winter) 0285 5611.

Country Life Museum, Sandy Bay, Exmouth, Devon EX8 5BU. Telephone: 0395 274533 or 039287 3230.

Craven Museum, Town Hall, High Street, Skipton, North Yorkshire BD23 1AH. Telephone: 0756 4079.

Cricket St Thomas Wildlife Park, Chard, Somerset TA20 2DD. Telephone: 046030 755.

Dairy Land, Tresillian Barton, Summercourt, Newquay, Cornwall TR8 5AA. Telephone: 087251 246.

Dorset County Museum, High West Street, Dorchester, Dorset DT1 1XA. Telephone: 0305 62735.

Easton Farm Park, Easton, Woodbridge, Suffolk IP13 0EQ. Telephone: 0728 746475.

Elvaston Working Estate Museum, Elvaston Castle, Elvaston, Derby DE7 3EP. Telephone: 0332 73799.

Finch Foundry Trust and Sticklepath Museum of Rural Industry, Sticklepath, Okehampton, Devon. Telephone: 0837 84352.

Folk Museum of West Yorkshire, Shibden Hall, Halifax, West Yorkshire HX3 6XG. Telephone: 0422 52246.

Gladstone Court Museum, Biggar, Lanarkshire ML12 6DN. Telephone: 0899 21050.

The Great Barn, Avebury, Marlborough, Wiltshire SN8 1RF. Telephone: 06723 555.

Guernsey Folk Museum, Saumarez Park, Câtel, Guernsey. Telephone: 0481 55384.

Gwent Rural Life Museum, The Malt Barn, New Market Street, Usk, Gwent NP5 1AU. Telephone: 02913 3777 or 063349 315.

Hampshire Farm Museum, Manor Farm, Brook Lane, Botley, Hampshire SO3 2ER. Telephone: 04892 87055.

Herefordshire Rural Heritage Museum, Doward, Symonds Yat, Ross-on-Wye, Herefordshire HR9 6DZ. Telephone: 0600 890474.

Highland Folk Museum, Duke Street, Kingussie, Inverness-shire PH21 1JG. Telephone: 05402 307.

The Irish Agricultural Museum, Johnstown Castle, Wexford, Republic of Ireland. Telephone: 053 42888.

The Iron Mills, A. Morris and Sons (Dunsford) Ltd, The Iron Mills, Dunsford, Exeter, Devon. Telephone: 0647 52352.

Killiow Country Park, Killiow, Kea, Truro, Cornwall TR3 6AG. Telephone: 0872 72768.

Lackham Agricultural Museum, Lackham College of Agriculture, Lacock, Chippenham, Wiltshire SN15 2NY. Telephone: 0249 656111.

Ladycroft Farm Museum, Elchies, Archiestown, Moray AB3 9SL. Telephone: 03406 274.

Lanreath Farm and Folk Museum, Churchtown, Lanreath, Looe, Cornwall PL13 2NX. Telephone: 0503 20321.

Launceston Rural Museum, Tremeale, South Petherwin, Launceston, Cornwall PL15 7JG. Telephone: 0566 2913.

Mary Arden's House, Wilmcote, Stratford-upon-Avon, Warwickshire. Telephone: 0789 293455.

Melton Carnegie Museum, Thorpe End, Melton Mowbray, Leicestershire. Telephone: 0664 69946.

Michelham Priory, Upper Dicker, Hailsham, East Sussex BN27 3QS. Telephone: 0323 844224.

Museum of East Anglian Life, Abbotts Hall, Stowmarket, Suffolk IP14 1DL. Telephone: 0449 612229.

Museum of English Rural Life, The University, Whiteknights, Reading, Berkshire RG6 2AG. Telephone: 0734 318660.

Museum of Kent Rural Life, Cobtree Manor Park, Lock Lane, Sandling, Maidstone, Kent ME14 3AU. Telephone: 0622 63936.

Museum of Lakeland Life and Industry, Abbott Hall, Kendal, Cumbria LA9 5AL. Telephone: 0539 22464.

Museum of Lincolnshire Life, The Old Barracks, Burton Road, Lincoln LN1 3LY. Telephone: 0522 28448.

Naseby Battle and Farm Museum, Purlieu Farm, Naseby, Northampton. Telephone: 0604 740241.

Norfolk Rural Life Museum, Beech House, Gressenhall, Dereham, Norfolk NR20 4DR. Telephone: 0362 860563.

Norris Museum, The Broadway, St Ives, Huntingdon, Cambridgeshire PE17 4BX. Telephone: 0480 65101.

North Cornwall Museum and Gallery, The Clease, Camelford, Cornwall PL32 9PL. Telephone: 0840 212954.

Old Kiln Agricultural Museum, Reeds Road, Tilford, Farnham, Surrey GU10 2DL. Telephone: 025125 2300.

Oxfordshire County Museum, Fletcher's House, Woodstock, Oxfordshire OX7 1SN. Telephone 0993 811456.

Park Farm Museum, Milton Abbas, Blandford Forum, Dorset DT11 0AX. Telephone: 0258 880216.

Priest's House Museum, 23-25 High Street, Wimborne Minster, Dorset BH21 1HR. Telephone: 0202 882533.

Rutland County Museum, Catmos Street, Oakham, Rutland, Leicestershire LE15 6HW. Telephone: 0572 3654.

Ryburn Farm Museum, Ripponden, Sowerby Bridge, West Yorkshire HX4 4DF. Telephone: 0422 54823 or 52334.

Ryedale Folk Museum, Hutton-le-Hole, York YO6 6UA. Telephone: 07515 367.

Scolton Manor Museum, Scolton, Spittal, Haverfordwest, Dyfed SA62 5QL. Telephone: 043782 328.

sh Agricultural Museum, Royal Highland Showground, Ingliston, Newbridge, Edinburgh
elephone: 031-333 2674.

ftesbury Local History Museum, Gold Hill, Shaftesbury, Dorset. Telephone: 0747 2157 o
3426.

omerset Rural Life Museum, Abbey Farm, Chilkwell Street, Glastonbury, Somerset BA6 8DB.
Telephone: 0458 32903.

Stacey Hill Museum, Southern Way, Wolverton, Milton Keynes, Buckinghamshire MK12 5EJ.
Telephone: 0908 316222.

Staffordshire County Museum, Shugborough, Stafford ST17 0XB. Telephone: 0889 881388.

Stewartry Museum, St Mary Street, Kirkcudbright. Telephone: 0557 30797.

Swaledale Folk Museum, Reeth Green, Reeth, Richmond, North Yorkshire DL11 6QT.
Telephone: 0748 84373.

Temple Newsam Home Farm, Leeds, West Yorkshire LS15 0AD. Telephone: 0532 645535.

Ulster American Folk Park, Camphill, Omagh, County Tyrone, Northern Ireland BT78 3QY.
Telephone: 0662 3292 or 3293.

Ulster Folk and Transport Museum, Cultra Manor, Holywood, County Down, Northern Ireland
BT18 0EU. Telephone: 0232 428428.

Upminster Tithe Barn Agricultural and Folk Museum, Hall Lane, Upminster, Essex RM14 1AU.
Telephone: 04024 47535.

Upper Dales Folk Museum, Station Yard, Hawes, North Yorkshire DL8 3NP. Telephone: 09697
494.

Welsh Folk Museum, St Fagans, Cardiff, South Glamorgan CF5 6XB. Telephone: 0222 569441.

Weston Park, Shifnal, Shropshire. Telephone: 095276 207.

Wimpole Home Farm, Arrington, Royston, Hertfordshire SG8 0BW. Telephone: 0223 207257.

Wolvesnewton Model Farm, Folk and Craft Centre, Wolvesnewton, Chepstow, Gwent. Telephone:
02915 231.

Yorkshire Museum of Farming, Murtonpark, Murton, York YO1 3UF. Telephone: 0904 489966.

The 1949 S-SL sugar-beet lifter-loader by Minns Manufacturing Company of Oxford. It topped, lifted and delivered the beet in one operation and could also be used on potatoes.